WHAT ARE EXOPLANETS?

Space Science Books Grade 4 | Children's Astronomy & Space Books

First Edition, 2020

Published in the United States by Speedy Publishing LLC, 40 E Main Street, Newark, Delaware 19711 USA.

© 2020 Baby Professor Books, an imprint of Speedy Publishing LLC

All rights reserved.

Without limiting the rights under the copyright reserved above, no part of this publication may be reproduced, stored in or introduced into a retrieval system, or transmitted, in any form, or by any means (electronic, mechanical, photocopying, recording, or otherwise), without the prior written permission of the copyright owner.

All images in this book have been reproduced with the knowledge and prior consent of the artists concerned, and no responsibility is accepted by producer, publisher, or printer for any infringement of copyright or otherwise arising from the contents of this publication.

Baby Professor Books are available at special discounts when purchased in bulk for industrial and sales-promotional use. For details contact our Special Sales Team at Speedy Publishing LLC, 40 E Main Street, Newark, Delaware 19711 USA. Telephone (888) 248-4521 Fax: (210) 519-4043. www.speedybookstore.com

10 9 8 7 6 * 5 4 3 2 1

Print Edition: 9781541953307
Digital Edition: 9781541956308

See the world in pictures. Build your knowledge in style.
www.speedypublishing.com

Table of Contents

"Exo" Means "Outside"..6
Early Astronomy and Exoplanets................................12
The Theories of Ancient Astronomers........................20
The Discovery of the First Exoplanet..........................26
A World of Exoplanets..32
Challenges to Finding Exoplanets...............................37
What Is the Radial Velocity Method?..........................41
What Is the Transit Method?..45
A Combined Method..48
What is the Closest Exoplanet to Earth?....................51
Going into Space to Study Space...............................57
Can We Visit an Exoplanet?..62
Summary..68

We know that our solar system, the Milky Way, has eight planets—Mercury, Venus, Earth, Mars, Saturn, Jupiter, Neptune, and Uranus—orbiting around the Sun, but these planets are not the only planets in the entire Universe. If you were to look up at the sky on a clear night, however, you would see billions of stars. Each one of these stars could, potentially, have one or more planets orbiting around it. Astronomers use the term "exoplanets" to describe planets that are located outside of the Milky Way solar system. In this book, we will take a closer look at exoplanets to learn about what they are, how astronomers have discovered some of them, and where they are located. Let's get started.

The Solar System

As a prefix, "exo", which was derived from a Greek word, means "outside, exterior, or outer". This means that the term "exoplanet" translated to mean planets that are located outside the Milky Way.

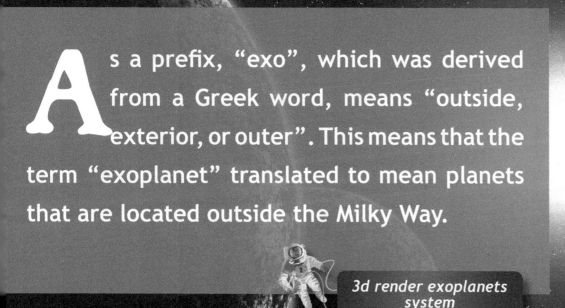
3d render exoplanets system

These are planets that orbit around a star different from our Sun. Like the planets in our own solar system, exoplanets can come in a variety of forms, from small, rocky planets to giant, gassy ones.

Exoplanet orbiting a star

Some are extremely hot while others are frozen worlds.

A 3D illustration of an exoplanet covered with ice and snow

Even their orbits are unique. Astronomers know of some exoplanets that have such small, tight orbits around their Suns that their years last only a few days on Earth. Others have much longer orbits. Some even orbit around two Suns.

Some exoplanets' orbit their Suns in just a few Earth days

Early Astronomy and Exoplanets

Humans have probably been wondering whether there are planets outside of the Milky Way for thousands of years.

An astronomer in the night with telescope

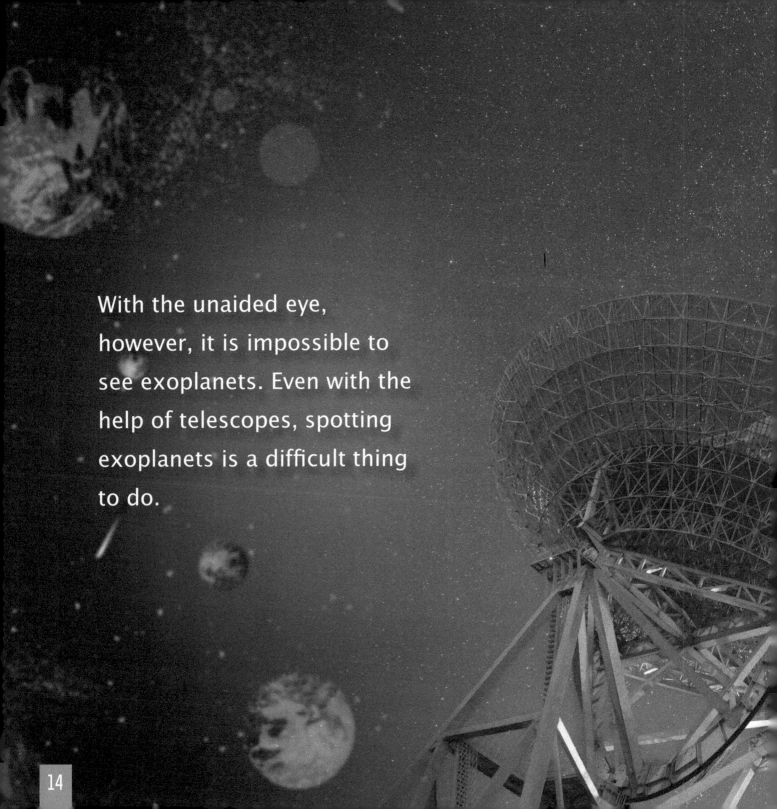

With the unaided eye, however, it is impossible to see exoplanets. Even with the help of telescopes, spotting exoplanets is a difficult thing to do.

Radio Astronomical Telescope at Astronomical Observatory, Beijing, China

The brightness of the star around which the exoplanet orbits hides the exoplanets from view. Still, many people felt certain that there must be additional planets in the universe.

This was a common theme in science fiction novels from as far back as the 1800s. It was also featured on television shows, such as Star Trek in the 1960s, and movies, including Star Wars in the 1970s.

While people speculated about the existence of exoplanets, scientists didn't discover one until recently.

Exoplanets were discovered only recently

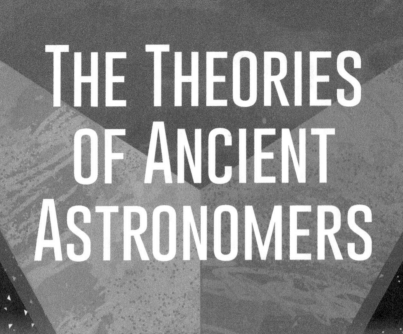

The Theories of Ancient Astronomers

An Italian philosopher named Giordano Bruno, who lived in the 16th century, piggybacked off the ideas of Copernicus.

Giordano Bruno

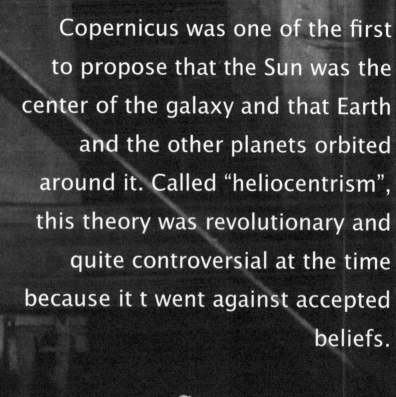

Copernicus was one of the first to propose that the Sun was the center of the galaxy and that Earth and the other planets orbited around it. Called "heliocentrism", this theory was revolutionary and quite controversial at the time because it t went against accepted beliefs.

Nicolaus Copernicus

Bruno suggested that the stars in the sky were all like the Sun and therefore could have planets encircling them just as the Earth and other planets orbit the Sun.

Giordano Bruno Statue, Campo de' Fiori, Rome

Isaac Newton wrote about this concept in his 18th century work Principia.

 Newton - Principia

The Discovery of the First Exoplanet

Using advance telescopes and equipment, astronomers first detected what they thought could be an exoplanet in 1988. It took a few years, until 1992, to confirm that the discovery was, indeed, an exoplanet.

Astronomical observatory Southold, New York, USA

It was named HIP 65426b and was found orbiting a pulsar, which scientists named PSR B1257+12.

HIP 65426b

PSR B1257+12

A few years later, another exoplanet was observed orbiting around the star, 51 Pegasi. This discovery, made in 1995, was an exciting moment in the scientific community and led to more observations into the space outside of the Milky Way.

51 Pegasi

31

A World of Exoplanets

Since that first exoplanet was discovered in 1988, astronomers have found many more. In fact, they have confirmed the existence of 4,126 exoplanets so far.

An artist's illustration of exoplanets orbiting around their host stars

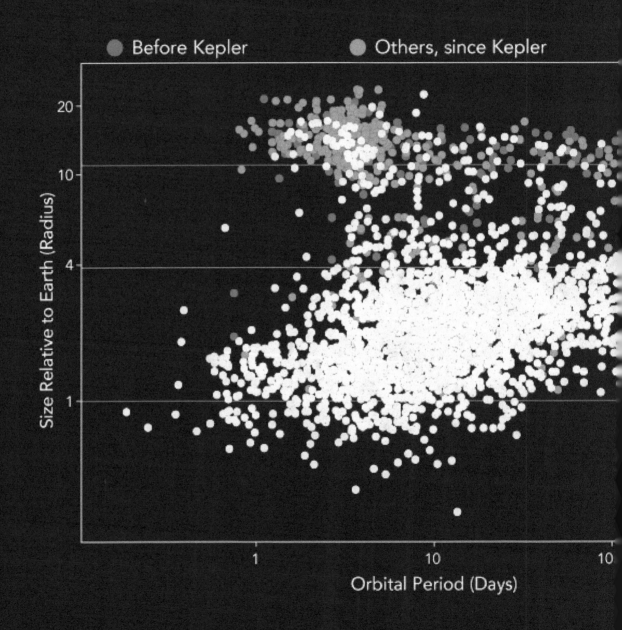

Kepler

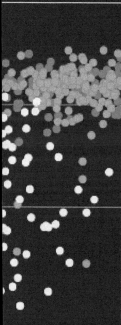

Jupiter

Neptune

Earth

These exoplanets are located in more than three thousand different solar systems. Like the Milky Way, 671 of these solar systems have more than one exoplanet orbiting their Suns.

Comparison of exoplanet discoveries before and after the use of the Kepler space telescope

1000

As of December 14, 2017

New exoplanets are still being discovered

With new discoveries being made all the time, it is expected that the number of stars in the universe will rapidly increase.

Challenges to Finding Exoplanets

It is not easy to spot exoplanets. First, they are very far away. They are also located so close to their Suns, that the brightness of those Suns obscures the exoplanets. For these reasons, only a few of the exoplanets have been discovered via direct observation. Most of the exoplanets that have been discovered were through indirect methods.

Exoplanets were discovered using either direct or indirect methods

Astronomers have found others by using either the radial velocity method, also called the wobble method, or the transit method.

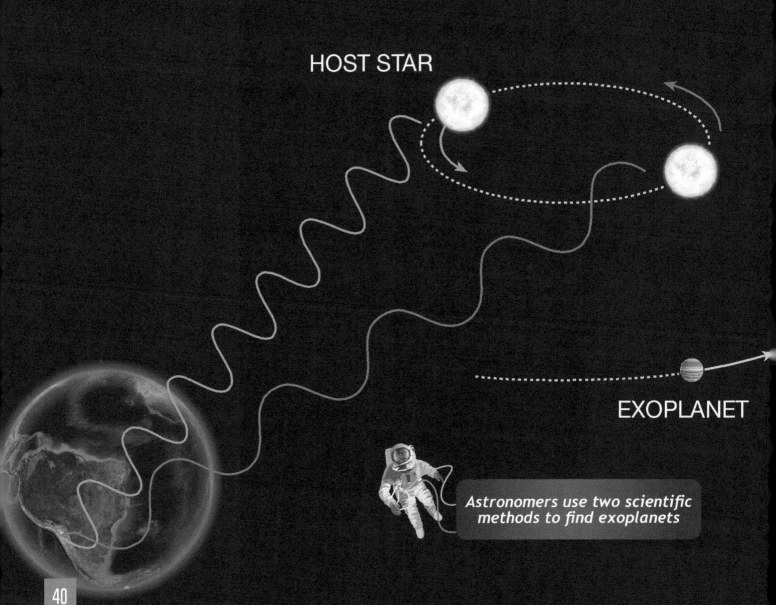

HOST STAR

EXOPLANET

Astronomers use two scientific methods to find exoplanets

What Is the Radial Velocity Method?

Astronomers can note the presence of exoplanets by making careful observations of the stars, looking for slight movements. The radial velocity method uses the tiny sways of a star as it is being pulled by an orbiting body. The gravitational force of the orbiting exoplanets exerts a tug on the star, making it sway or wobble. This is why the radial velocity method is also called the wobble method.

This artist's illustration demonstrates the "wobble," or radial velocity, technique for finding planets

Unseen planet

Doppler Shift
due to Stellar Wobble

Scientists also observe how much the star wobbles because this is an indicator of the size and mass of an exoplanet. The radial velocity method was how the first exoplanet was discovered. In fact, this method is responsible for the discovery of at least 700 exoplanets to date.

What Is the Transit Method?

Most of the known exoplanets have been found using the transit method. The transit method required searching for tiny shadows that are cast on stars by passing objects. The light emitting from a star decreases ever so slightly when an object crosses in front of it, an event that astronomers call a transit.

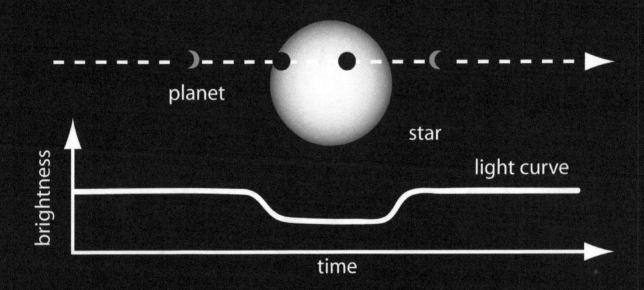

Transit method

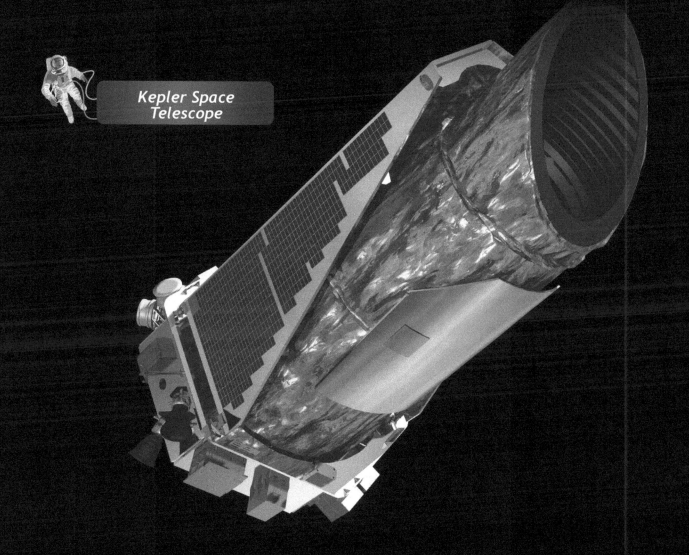

Kepler Space Telescope

In many cases, the observation of transits occurs in space, instead of on Earth. The Kepler space telescope, which NASA launched in 2009, has discovered nearly 2,700 exoplanets during its mission.

A Combined Method

Since there are two ways to detect the presence of exoplanets, astronomers often use both methods together. For example, they may suspect that an exoplanet is the cause of a star wobble, so they will continue to observe this star in hopes of observing a transit that will confirm that an exoplanet is orbiting the star.

Kepler Space Telescope

Scientists use both methods to confirm the existence of exoplanets

This combined method works like a checks and balance system to help scientists confirm that an exoplanet is in that location. It also helps astronomers learn additional information about the exoplanets, such as size and mass.

What is the Closest Exoplanet to Earth?

The closest exoplanet to Earth, at least that we know of so far, is Proxima Centauri b, a planet that orbits the star that is located closest to the Milky Way. This star is known as Proxima Centauri.

Artist's impression of Proxima Centauri b

The exoplanet orbiting this star was confirmed in August of 2016 when it was discovered using the radial velocity method by the European Southern Observatory.

ESO's Paranal Observatory in Chile

Proxima Centauri b is about 4.2 lightyears away from Earth, the equivalent of 25 trillion miles. This exoplanet orbits its Sun closely; its year is equal to a little bit more than 11 Earth days. Although we do not yet have the capability to send spaceships to Proxima Centauri b at this point, it is feasible that humans could send robotic probes to visit this exoplanet at some time in the future.

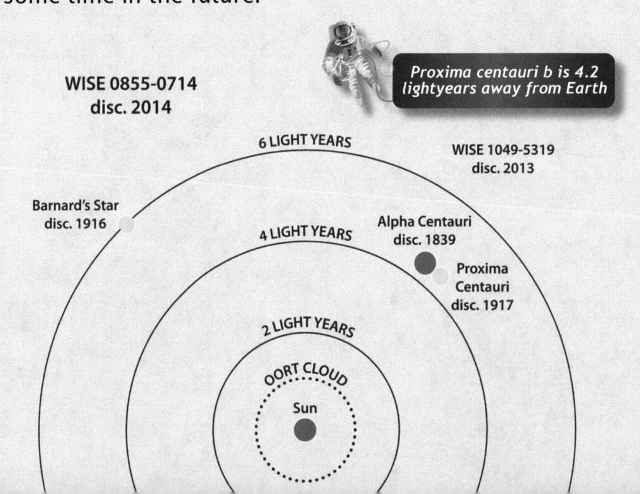

Proxima centauri b is 4.2 lightyears away from Earth

Going into Space to Study Space

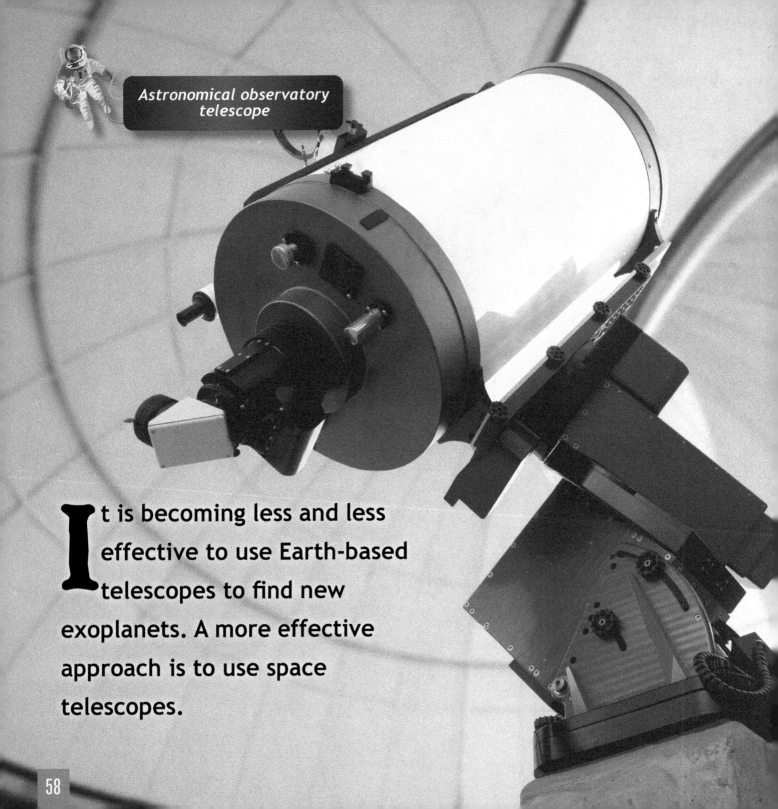

Astronomical observatory telescope

It is becoming less and less effective to use Earth-based telescopes to find new exoplanets. A more effective approach is to use space telescopes.

The Kepler space telescope that was launched into space in March of 2009 proved this by sending back data that led to the discovery of more than 2,700 exoplanets. The Kepler space telescope, however, ended its mission in October of 2018 when its fuel supply ran out.

Kepler Telescope

Artist's rendering of the Transiting Exoplanet Survey Satellite (TESS)

NASA replaced Kepler with the Transiting Exoplanet Survey Satellite, or TESS, which was launched in April of 2018. TESS is much larger and more powerful than Kepler.

Another space telescope, the James Webb Space Telescope, will be launched in 2020. This space telescope is even more powerful than TESS.

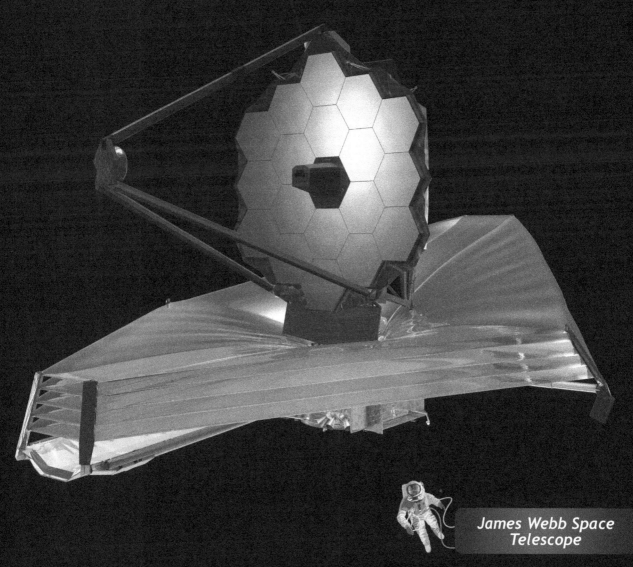

James Webb Space Telescope

Can We Visit an Exoplanet?

All of the exoplanets that have been discovered so far, including Proxima Centauri b, are really, really far away, as much as several thousand light-years away. We do not yet have the technology to send astronauts to explore these exoplanets.

Exoplanets are several thousand light-years away from Earth

It will likely be a few hundred years before we can make that a reality. The observations that astronomers are currently doing on these exoplanets, however, are discovering information that will help us narrow down the list of exoplanets to the ones that may be suitable for habitation.

A 3d illustration of Exoplanets

A 3D illustration of an exoplanet's landscape

Unlike the TV shows, movies, and science fiction novels, not all exoplanets can support life. The conditions need to be just right to make an exoplanet a place that humans could safely stay.

Summary

As the name implies, exoplanets are planets that are located outside of the Earth's solar system. So far, astronomers have discovered more than 4,100 exoplanets and it seems likely that there are many, many more. The vast majority of exoplanets have been discovered by indirect methods, either the radial velocity method or the transit method. The Kepler space telescope found the vast majority of the known exoplanets, but this space telescope is no longer operational. Fortunately, newer, more powerful telescopes are scanning the universe and will undoubtedly discover more new worlds.

There is a huge expanse of space beyond our solar system. If you enjoy learning about the planets in the Milky Way, you will like to learn about the exoplanets that have been discovered and the possibilities that these newly found worlds hold.

Visit

www.BabyProfessorBooks.com

to download Free Baby Professor eBooks and view our catalog of new and exciting Children's Books

CPSIA information can be obtained
at www.ICGtesting.com
Printed in the USA
BVHW011651181022
649735BV00021B/200